Bay Area Biotechnology 2016

Topical Conference at the 2016 AIChE Annual Meeting

San Francisco, California, USA
13 - 18 November 2016

ISBN: 978-1-5108-3425-5

Printed from e-media with permission by:

Curran Associates, Inc.
57 Morehouse Lane
Red Hook, NY 12571

Some format issues inherent in the e-media version may also appear in this print version.

Printed by Curran Associates, Inc. (2017)

AIChE
120 Wall Street, FL 23
New York, NY 10005-4020

Phone: (800) 242-4363
Fax: (203) 775-5177

www.aiche.org

Additional copies of this publication are available from:

Curran Associates, Inc.
57 Morehouse Lane
Red Hook, NY 12571 USA
Phone: 845-758-0400
Fax: 845-758-2633
Email: curran@proceedings.com
Web: www.proceedings.com

TABLE OF CONTENTS

475260 Next-Generation Testing Platform for Strain Optimization, Screening, and Fermentation Testing

Abstract:

Amyris is a renewable products company that engineers yeast to convert sugar into valuable chemicals. We have successfully produced four isoprenoid molecules at commercial scale, replacing petroleum, animal or plant-sourced chemicals with reliable, sustainable alternatives. Underlying this commercial success is an automated pipeline for the construction, testing, and deployment of our yeast strains that has made each strain development timeline shorter with each successive molecule. We are now in the midst of expanding our footprint on the metabolic map into chemical targets outside isoprenoids, and consequently, expanding the capabilities of our strain construction pipeline. This talk will focus on our strain testing technologies, and describe the next-generation testing tools we are developing to enable testing of strains targeting diverse chemical space and multiple global markets.

475263 Harnessing Biological Complexity and Diversity through Metabolic Engineering: The Manus Bio Approach

Abstract:

Specialized metabolites represent a class of compounds with unique and complex chemistries and which possess a myriad of commercial applications, from pharmaceuticals and food products to industrial chemicals. Despite their intrinsic value, however, sourcing remains a bottleneck to more widespread use due to their low abundance in nature and the inherent difficulties in applying metabolic engineering and industrial biotechnology approaches to produce these complex natural chemistries in microbial hosts. At Manus Bio, we focus on addressing the challenges of microbial fermentation of plant secondary metabolites through the combined application of our proprietary Multivariate Modular Metabolic Engineering (MMME) and Pathway Integrated Protein Engineering (PIPE) technologies for optimizing heterologous enzymes and pathways. With this approach, we have successfully demonstrated the potential of using a single microbe to catalyze several complex chemical reactions, including polymerization, cyclization, multiple oxidations, and up to six glycosylations, all while producing the target metabolite at greater than 95% purity. In this presentation, we will highlight some of the important insights and guiding principles used to engineer complex biochemical pathways in simple bacterial systems, such as *Escherichia coli*.

475265 The Emerging Organism Engineering Industry

Abstract:

There is an emerging demand for sourcing plant-derived extracts (nutraceuticals, flavors, fragrances, sweeteners, etc.) from engineered microbes. While recent advances in synthetic biology and metabolic engineering provide feasible approaches to engineering such organisms, commercial success for developing these â€œculturedâ€ ingredients presents specific challenges. Unlike biofuels, where efforts can be focused on one particular molecule given the enormous market size, cultured ingredients require developing different organism lines in a rapid and low cost fashion. This requires a scalable solution for bio-manufacturing of organisms, which is provided by our state of the art foundry that continues to grow. I will describe how organism development at Ginkgo leverages our foundry to accelerate the design/build/test using specific examples. In particular, I will highlight the value of combining computer-aided engineering software tools, cheap gene synthesis and high resolution-accurate mass LCMS to develop engineered microbes. Finally, Iâ€™ll touch on how our improvements in manufacturing organisms lend to opportunities outside of cultured ingredients.

475269 General Purposing Synthetic Biology: a Biofoundry in a Box

Abstract:

What could you build with unlimited DNA writing? Productivity of DNA synthesis continues to increase at exponential rates. The oligonucleotide synthesis market alone is forecast to reach nearly $2B USD by 2020 with an annual growth rate in excess of 10%. This increase in oligo synthesis capacity and the associated reduction in cost is transforming synthetic biology; cheap oligo synthesis will remake the impossible to the new routine. Muse's foundational technology is a proprietary synthetic biology platform that scales directly with multiplex oligo synthesis, enabling designer DNA construction at 1000X lower cost than the current state of the art. To date, Muse has demonstrated the ability to generate and track over 50,000 designer mutants simultaneously. Applications of this technology include sequence to activity mapping for site saturation mutagenesis, protein and pathway engineering, and strain reconstruction of complex evolved strains to name a few. Gene editing capabilities have been demonstrated in various industrially relevant organisms, including applications across biotechnology. Further, this technology is being automated in the form of a bench-top device that allows a single user to perform a broad range of molecular biology endeavors at rates 1000X beyond the current state of the art.

475259 Making Silk without Silkworms: Using Industrial Biotechnology to Make Performance Protein-Based Fibers

Abstract:

Bolt Threads is a Bay Area startup that believes that the combination of molecular biology and materials science can revolutionize how people think about and use natural materials. Bolt's materials can dramatically improve the performance and environmental impact of a broad range of consumer and industrial materials. Specifically, Bolt is developing the next generation of sustainable performance apparel by leveraging the latest advances in industrial biotechnology, automation, and materials science. Derived from our knowledge of spider silk protein, the company has developed a high-throughput platform to generate thousands of different silk proteins, all from a scalable microorganism host. Using industrial biotechnology processes and fiber spinning, Bolt is able to make commercial volumes of protein and fiber. Precision control from primary protein sequence through fiber extrusion allows the company to produce materials that combine the benefits of both natural and synthetic fibers, and to go beyond even what is currently available in nature and the marketplace. Leveraging the unique properties of its fibers, Bolt Threads will be delivering its own line of consumer apparel made with its Engineered Silk(TM) as well as working with partners across industry verticals.

475548 The Development of Platform-Based Technologies for the Optimization of Sustainably Produced Chemicals

Abstract:

Genomatica has established an integrated computational/experimental bioengineering platform to design, create, and optimize novel high-producing organisms and bioprocesses. Here we present the use of our platform to develop *E. coli* strains for the production of the industrial chemical 1,4-butanediol (BDO) from carbohydrate feedstocks. These efforts have culminated in commercial scale production of over 5M lbs of BDO from dextrose. BDO is a chemical intermediate produced commercially through energy-intensive petrochemical processes from hydrocarbon feedstocks and is used in a variety of products including automotive, electronics and apparel. Therefore, this product represents an opportunity to make a significant impact on the replacement of traditional petrochemical processes with benign bioprocesses using renewable feedstocks.

We are continuing to enhance our technology platform, enabling faster development of new processes with increased chance of success. Here we describe the application of cell-free transcription-translation (TX-TL), towards improving the design-build- test-learn cycle for the construction of a high-performing microorganism capable of producing BDO and other industrial chemicals from carbohydrates. We used systems biology approaches including 13 C-flux analysis, proteomics, and metabolomics to characterize TX-TL applied to different engineered metabolic pathways, including BDO, and compared the results to the analogous in vivo processes. These studies provide critical insight into the active metabolism of TX-TL and mechanisms of redox and energy regeneration. Learnings from these studies were applied towards improving the TX-TL platform for the production of intermediate and specialty chemicals. The presentation will highlight the successful integration of the TX-TL technology into our rational and high-throughput strain engineering platform, and will show the potential advantages for using cell-free systems towards the optimization of sustainably produced chemicals.

478302 Imagine, Design, Create â€“ Cloud Based Software for Biological Design

Abstract:

Over the past 10 years the ability to sequence, edit, and build life has grown exponentially. Software in this space has not kept pace with the rapid pace of biological innovation resulting in a tremendous gap between the modeling and manufacturing of biological systems. The Autodesk BioNano team works to bring design solutions to biology, making the interaction of scientists with software more seamless and efficient. Our goal is to allow scientists to focus on what matters, the science. Autodesk has experience in designing and displaying complex, man-made mechanical structures in the cloud. It is time to apply this knowledge to biology to show how proteins and molecules interact to create life. Autodeskâ€™s BioNano group hopes to empower scientists to Imagine, Design, and Create a better world.

475268 Technology Transfer and Scale-Down Models for Mammalian Cell Culture Manufacturing   the Process Knowledge Questions, and the Practical Ones

Abstract:

Fed-batch processing using mammalian cell culture is the most widely used technology for large-scale manufacturing of monoclonal antibodies. Process optimization for the production of monoclonal antibodies is a continued process, potentially involving multiple process generations, and manufacturing sites.

Manufacturing Science groups are a key part of process optimization, as part of their responsibilities is to perform technology transfer from development groups to commercial manufacturing sites, or from a commercial manufacturing site to another. Technology transfer may involve the utilization of lab-scale systems to build process knowledge and to understand performance before manufacturing initiates at large scale. Once commercial scale production is in place, scale-down models are used to monitor and provide real time troubleshooting.

This presentation will focus on the establishment of a scale-down model to support the technology transfer of a new generation of a commercial product to Genentech Vacaville, a multiproduct commercial manufacturing site with the capability of producing monoclonal antibodies in large quantities. Work was performed to optimize the scale-down model, while at the same time provide answers to questions related to manufacturability (e.g., differences in process and impact to run rate, and limitations due to equipment capability). Scale-down model optimization tested the impact of process parameters on performance and product quality, while scale-down model utilization provided answers that enable process fit at large scale.

475264 Metabolic Engineering of *Saccharomyces Cerevisiae* for Increased Production of Opioids

Abstract:

Opioids are the primary drugs used in Western medicine for pain management and palliative care. Farming of opium poppies remains the sole source of these essential medicines despite diverse market demands and uncertainty in crop yields due to weather, climate change, and pests. Researchers at Stanford University recently demonstrated production of the opioid compounds thebaine and hydrocodone from sugar in engineered yeast strains. This milestone achievement combined enzyme discovery, enzyme engineering, and strain optimization to assemble a functional pathway containing over 20 different enzyme activities from plants, mammals, bacteria, and the native yeast host. At Antheia, Inc. we are further developing yeast strains and fermentation processes to translate this proof-of-concept work into a commercially viable venture. The result will be a sustainable, lower cost, and more secure and stable supply of opioid pharmaceutical ingredients.

475253 Mining for Therapeutics from the Microbiome

Abstract:

Bacteria that live in and on us (i.e., the microbiome) make thousands of molecules that have drug-like properties. There is a new revolution afoot to mine for metagenomic functions of interest, clone and express such genes, and test the function of the bioactive (protein/peptide, etc.). in appropriate in vitro and in vivo models. Since many of these bioactives never previously been purified, characterized, and tested in therapeutic doses, this has potential to open up entirely new biology which can be exploited for finding new therapies for many diseases. Building a scalable microbiome drug discovery platform based on this premise requires overcoming key big data and engineering challenges that will be the main focus of this presentation.

475254 Antibiotic Probiotics Reduce Salmonella Enterica in Pre-Harvest Poultry

Abstract:

Chicken has become the most popular type of meat in the world. In the US, per capita consumption has doubled to nearly 90 lbs/year since 1980. The value of chicken broilers produced during 2014 was $32.7 billion, up 6 percent from 2013.

Chicken is also the leading cause of salmonella food poisoning. Despite the vast arsenal of technologies employed to control salmonella in poultry products, these products are the most significant sources of Salmonella enterica that sickens people. In the US, Salmonella enterica is the leading cause of foodborne disease outbreaks, with approximately 1 million infections annually, and calculated health care and lost productivity costs exceeding $3 billion every year. These disconcerting numbers may increase after the FDA-directed phasing out of the use of medically important antimicrobials in food animals for production purposes.

Herein, we report on a new antibiotic technology that can reduce Salmonella enterica in poultry. We have engineered the probiotic strain E. coli Nissle 1917 to produce and secrete the antimicrobial peptide microcin J25. Employing synthetic biology techniques we engineered a DNA promoter for strong, constitutive expression of the peptide. In vitro tests demonstrate the efficacy of the system against Salmonella enterica. The reported animal experiments provide strong proof-of-concept results. Salmonella enterica clear the ceca of poultry faster than in negative control experiments. Approximately 97% lower salmonella carriage is measured in the treated group, by end of the experiment, 14 days post-challenge, compared to the untreated group.

Probiotic bacteria are generally regarded as safe to consume, are bile-resistant and can plausibly be modified to produce many out of the thousands of antimicrobial peptides now known. The reported systems may provide a foundation for platforms to launch antimicrobials against any one of many gastrointestinal tract pathogens, including ones that are multi-drug resistant.

475392 Use of Synthetic sgRNA to Improve CRISPR Editing Efficiency

Abstract:

CRISPR has made genome editing accessible for a wide range of cell-types. However, obtaining consistent editing efficiencies remains a challenge. Synthego has developed novel RNA synthesis technology to produce 100-mer sgRNA for CRISPR at a practical scale and price. We demonstrate that Synthego sgRNA produces consistent and improved genome editing.

481290 From Mind to Molecule: Building the Information Technology Backbone for Synthetic Biology

Abstract:

Biology is now an information science. Our ability to read, write and edit biology has undergone a radical transformation in just the last few years. At TeselaGen we are building software systems that help scientists and managers keep up with our rapidly expanding ability to sequence, synthesize and edit biology. In collaboration with Dow, TeselaGen is building an enterprise platform for synthetic biology that is secure, scalable and expandable and powers rapid, precise and accurate DNA modification. With automatic protocol generation using patented *j5* technology from Berkeley Lab and the best available CRISPR-Cas based techniques, TeselaGen is advancing in our ability to design and build in the biological realm in much the same way that electronic design automation enabled design and build in the electronic realm. TeselaGen is an early stage company that makes its home in the software hub of San Francisco, CA. Our customers include medical, industrial and agricultural companies from around the world that leverage biotechnology to make medicines, enzymes, bio-based chemicals, and help farmers with plants that can better withstand pests and environmental stress.

475309 The Power of Prediction: Rapid Optimization of Metabolic Pathways without High-Throughput Screening

Abstract:

The Salis Lab and De Novo DNA have developed an integrated computational-experimental pipeline that rapidly designs and optimizes many-protein genetic systems, such as metabolic pathways & networks, inside engineered micro-organisms, while requiring only a small number of characterization experiments. Our pipeline can be applied to any large genetic system to optimize its performance, using only two design-build-test-learn cycles. To do this, we've combined several predictive models, design rules, and optimization algorithms to map the genetic system's sequence-expression-activity relationship, predict its optimal protein expression levels, and prioritize protein engineering efforts. Our platform is web-accessible at http://www.denovodna.com/software, and has been used by over 6000 registered researchers to design over 100,000 synthetic DNA sequences. We present case studies to demonstrate the pipeline's process.

475294 Reimagine Sequence Space

Abstract:

The applications enabled by perfect DNA sequences from a highly scalable synthesis platform are limitless. With Twist Bioscience's innovative silicon-based platform, reaction volumes are miniaturized by a factor of 1000 enabling high quality DNA production at reduced price and turn-around time. Delivering DNA as genes, ready-to-use oligonucleotide pools or variant libraries simplifies experiments by circumventing the need to clone, allowing use of sequence space to be reimagined on a grander scale, ultimately accelerating discoveries.

475262 Rapid Optimization of Industrial Microbes Via High-Throughput Strain Engineering

Abstract:

Zymergen has built a technology platform for high-throughput engineering of a wide variety of microbes. We use this technology platform to build and test many design ideas and rapidly improve strain performance. In this presentation, an overview of our technology platform for strain design, strain construction and strain testing will be given. In addition, selected case studies will be discussed.

475251 Optimizing Protein Production Strains for Downstream Purification

Abstract:

Boston Mountain Biotech is working on simplifying the removal of host cell proteins (HCPs) that unfavorably bind during protein purification. Despite major advancements in biopharmaceutical manufacturing, protein purification remains one of the most costly and restrictive aspects of production. BMB seeks to directly address this challenge by developing the Lotus® manufacturing platform that identifies the HCPs that exert the greatest burden on downstream processing and then silences these genes in the production strain. This platform could be readily implemented across much of the biomanufacturing industry, with the potential to reduce costs to consumers in the ever-burgeoning healthcare market and ease the time and investment required to bring new therapeutics to market.

475258 Chemically Synthesized Modified Guide RNAs Enhance Crispr/Cas Genome Editing

Abstract:

Using an advancement in RNA synthesis technology (Dellinger et al. J. Am. Chem. Soc. 2011, 133, 11540â€“11556), we find it straightforward to chemically synthesize RNAs of more than 100 nt in length and to incorporate modified nucleotides at any position. Using this technology, we have demonstrated that chemical alterations to synthesized single guide RNAs (sgRNAs) enhance genome editing efficiency in human primary T cells and CD34+ hematopoietic stem and progenitor cells (Hendel et al. Nat. Biotech. 2015 doi:10.1038/nbt.3290). Here, we extend this work characterizing the performance of chemically modified sgRNAs for genome editing in additional cell types including human primary cells (e.g. Fibroblasts, hepatocytes) and induced pluripotent stem cells. The ability to chemically synthesize high quality sgRNAs in a scalable manner enables their widespread application in all phases from basic research to therapeutic use.

AUTHOR INDEX